Royal Navy and Army Helicopters of the 1970s and '80s

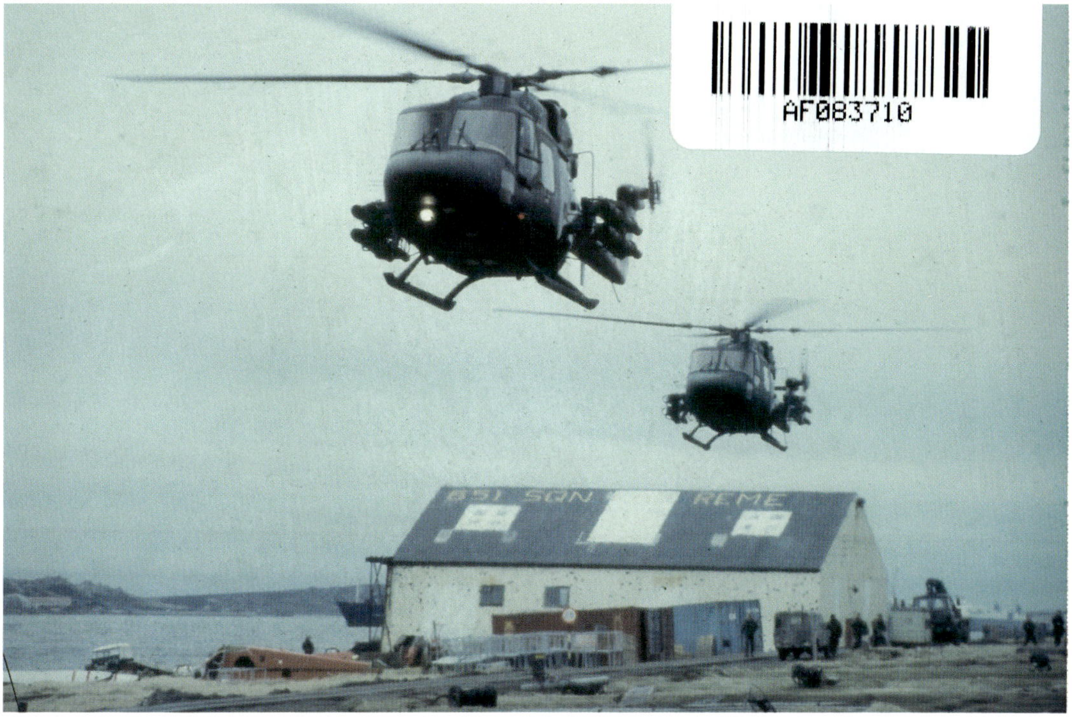

CHRIS GOSS

KEY
Books

Published by Key Books
An imprint of Key Publishing Ltd
PO Box 100
Stamford
Lincs PE9 1XQ

www.keypublishing.com

The right of Chris Goss to be identified as the author of this book has been asserted in accordance with the Copyright, Designs and Patents Act 1988 Sections 77 and 78.

Copyright © Chris Goss, 2023

ISBN 978 1 80282 259 5

All rights reserved. Reproduction in whole or in part in any form whatsoever or by any means is strictly prohibited without the prior permission of the Publisher.

Typeset by SJmagic DESIGN SERVICES, India.

Contents

Foreword 4

Introduction 5

Chapter 1   Lynx 6

Chapter 2   Sea King 21

Chapter 3   Wasp/Scout 35

Chapter 4   Wessex 49

Chapter 5   Whirlwind 67

Chapter 6   Minor Types 76

Appendix   Variants 94

Foreword

I am delighted to write this foreword for Chris Goss' latest book on Royal Navy and Army helicopters. I have always been passionate about flying and was determined to become a helicopter pilot in the Fleet Air Arm (FAA) ever since I was a small boy. With my father being in the Royal Navy, I was born in a married quarter at HMS Seahawk at RNAS Culdrose and remember being entranced by successive Air Days at Culdrose and HMS Heron at RNAS Yeovilton. It must have been when I was four years old and one of my first memories was watching a flying display by a formation of five Mk 1 Sea Vixens from 766 Naval Air Sqn called 'Fred's Five', led by Lt Cdr Peter Reynolds who was 'Fred'. It was this memory that stayed with me and made me determined to join the Navy and go on to earn my Wings in 1983.

With an FAA career starting with flying ASW Sea Kings and Lynx helicopters at sea as a relatively junior officer, I was able to return to flying much later as the Deputy Commander and then the Commander of the Joint Helicopter Command. At the time, our challenge was maintaining Joint helicopter air groups in two theatres of operation in Iraq and Afghanistan simultaneously and continuously, as well as deploying 16 Air Assault Brigade to Afghanistan three times. The mixed set of aircraft, flown by all three Services comprised Chinooks, Merlin and Pumas from the Royal Air Force, Sea Kings from the RN Commando Helicopter Force, and Lynx and Apaches from the Army Air Corps. The extraordinary game-changing effect that all these helicopters and their crews had in such challenging operational environments was demonstrated every day; saving lives by recovering the critically injured, deploying troops deep in enemy territory, supporting both Special Forces and major ground assaults, and conducting recces and airborne command and control. Their multirole flexibility and versatility proved to be the ultimate antidote to the asymmetric and hybrid warfare waged by the insurgents in both theatres.

This book of helicopters spanning over 20 years flown in the Royal Navy and Army reflects this unique 'force-multiplying' capability and adaptability perfectly. Chris Goss has produced a superb set of 180 photographs for enthusiasts, aviation specialists, current and former aircrew, and the community's extensive family of air engineers, weapons specialists, radio and electrical specialists, air traffic controllers, airfield specialists and flight deck handlers. What has made this book even more special for me is that having cross-checked my aircrew logbook with the airframe numbers of the aircraft in the photographs, I have found at least seven that I actually flew! I hope that you enjoy it as much as I have done.

Vice Admiral Sir Tony Johnstone-Burt KCVO CB OBE DL

Introduction

This book is the sixth in the series devoted to photographs of British combat aircraft of the 1970s and 1980s, the last two decades of the Cold War.

The first book in the series covered the English Electric Lightning and McDonnell Douglas Phantom fighter aircraft, the second the Blackburn Buccaneer and Avro Vulcan bombers, and the third looked at the SEPECAT Jaguar and Harrier ground-attack aircraft as well as the Sea Harrier. The fourth book looked at the Avro Shackleton, Hawker Siddeley Nimrod, English Electric Canberra and Fairey Gannet – the reconnaissance and airborne-early warning aircraft, while the fifth book reviewed the Boeing CH-47 Chinook, Westland Sea King, Westland Wessex and Westland Whirlwind Royal Air Force (RAF) helicopters.

This book presents those helicopters used by the Royal Navy (RN) and the British Army, namely the Westland Lynx, which was used by the RN and Army, and the Sea King, Wessex and Whirlwind, which were used by the RN. The Westland Wasp used by the RN and the Westland Scout used by the Army are also included. A final section covers lesser-known or lesser-appreciated smaller helicopters, such as the Aérospatiale Alouette and Bell Sioux used by the Army and the Westland Gazelle used by both forces.

The roles these helicopters undertook were varied and included Search and Rescue (SAR), support (SH), anti-tank (AT), anti-submarine warfare (ASW), airborne observation (AOP), commando support and airborne early warning (AEW) duties. Helicopters such as the Wasp and Wessex played an active role in the Falklands War.

Yet again, I would like to thank Bernd Rauchbach, for checking the captions for me, and Andy Thomas, for his continued advice and for being a fount of all knowledge on all things RN and Army despite being ex-RAF. Furthermore, I would like to recommend (and thank the organisers of) the Helicopter History website (https://www.helis.com/) as an excellent research tool for anything helicopter related. Finally, once more, this book is dedicated to the late David Howley, without whose generosity I would not have had access to so many photographs of aircraft he encountered, photographed and logged during and after his RAF career.

Chris Goss
Marlow 2023

Chapter 1
Lynx

Westland WG.13 XW389 was the fifth prototype of what would later become the Lynx and first flew in June 1974. It was seen at the Farnborough Air Show later the same year, where this photo was taken, painted a dull red. It appears to have continued trials work with Rolls-Royce and Westland, and by 1980, it was being used for ground instruction at the Royal Naval Engineering College (RNEC) in Manadon, Devon, after which it went to Royal Navy Air Station (RNAS) Yeovilton's fire dump. In 1996, it went to the Helicopter Museum at Weston-Super-Mare, where it is still in 2023.

Also seen at Farnborough in 1974 was a pre-production Lynx HAS.2 XX910, painted in Royal Navy (RN) colours and carrying a drill Mk 46 torpedo. It first flew in April 1974 and was delivered to the Ministry of Defence (MoD) the following year. Its claim to fame was that it performed a loop at the Farnborough Air Show in September 1976. It then went into service with the Royal Aircraft Establishment (RAE) and the Aircraft and Armament Experimental Establishment (A&AEE), before returning to the RAE at Farnborough. From there, it went into storage and was used for spares, but in January 2000 it went to the Helicopter Museum, where it remains today.

An interesting line up of front-line RN aircraft at RNAS Yeovilton, celebrating The Queen's Silver Jubilee in 1977. Nearest to the camera is Lynx HAS.2 XZ234, which was delivered to the RN in May 1977. It first flew with 700L Sqn in 1977 before going to 702 Sqn. It would later be converted to a HAS.3, but by 2010 it was being used for spares at Middle Wallop airfield, after which the remains were sold. It is now believed to be in private hands in Suffolk.

Two Lynx HAS.2s of 702 Sqn formatting with Dutch Navy counterparts. To the rear is the fourth production model, XZ229, which was delivered in 1976. It had a long career with 702, 815, 829 sqns before being converted to a HAS.3 in 1988. It then returned to 829 Sqn, after which it flew with 702 and 815 sqns. In 2011, it was used for spares before being sold in 2013. To the front is XZ231, which had a similar career to XZ229 but was sold to Pakistan in 1995.

Seen off HMS *Sirius* in 1980 is Lynx HAS.2 XZ231. It had first flown in October 1976.

Seen operating off an unidentified warship are Lynx HAS.2s XZ229 (airborne) and XZ232. Both first flew with 700L Sqn, and XZ232 later became a training airframe at Aberdeen Airport and is owned by RelyOn Nutec.

Another two Lynx HAS.2s of 700L Sqn. The letters 'EN' on the flight deck would indicate the ship is RFA *Engadine*. XZ233 first flew in 1977, flying with 700L, 702 and 815 sqns before being converted to a HAS.3 in 1986 and then a HAS.3S. It was stripped for spares and later scrapped in 2017.

Lynx AH.1 XX153 preparing for air transportation. This aircraft first flew in April 1972, and its claim to fame was that, between 20–22 June 1972, while being flown by Westland chief pilot Leonard Roy Moxham and flight test engineer Michael Ball, it set two Fédération Aéronautique Internationale world records for speed. Flying over a straight 15/25km course on 20 June, the Lynx averaged 321.74kph (199.92mph). Two days later, it flew a closed 100km circuit at an average speed of 318.50kph (197.91mph). It can be seen today at the Museum of Army Flying at Middle Wallop.

Seen at RNAS Yeovilton in June 1977 is Lynx HAS.2 XZ232 of 700L Sqn. This aircraft can be seen at Aberdeen Airport, where it is currently being used as a training aid.

Lynx HAS.2 XZ230 of 702 Sqn at RNAS Yeovilton in August 1979. This aircraft was delivered to 700L Sqn in October 1976, after which it joined 702 Sqn. It was subsequently upgraded to HAS.3 standard and HAS3.GSM and served with 815 and 829 sqns. It took part in Operation *Desert Storm* in 1990, flying with 815 Sqn off HMS *Cardiff*. On 2 November 1998, it was damaged beyond repair when serving with 815 Sqn on HMS *Newcastle* when the flight deck was engulfed by a massive wave. Despite being sold and then used as a film prop, it is currently believed to be back in RN service and being used for ground instruction at RNAS Yeovilton.

Lynx AH.1 XZ176 of the Lynx Conversion Flt (LCF) was delivered in November 1977 and is known to have flown with 651, 665 and 671 sqns of the Army Air Corps (AAC). It was upgraded to an AH.7 around 1998 and can be seen today at Camping Land Uit Zee, Wieringerwerf, The Netherlands.

A Lynx HAS.2, flanked by three Sea Harriers of 899 Sqn, at RNAS Yeovilton in August 1980.

Lynx HAS.2 XZ689 of 702 Sqn landing at Yeovilton in August 1981. Built in 1978, it was upgraded to HAS.3 standard and finally HMA.8 DSP and HMA.8 SRU (also known as a Super Lynx) standard. In June 2022, it was reported as being used for spares recovery at Hayward and Green Aviation Ltd, Woodmancote, West Sussex.

A dramatic photograph of two Lynx AH.1s at Port Stanley in the Falkland Islands in 1982. Each are armed with eight Tube-launched, Optically tracked, Wire-guided (TOW) anti-tank missiles.

Displaying at RNAS Yeovilton in July 1982 are three Lynx AH.1s of the LCF.

Seen at Royal Air Force (RAF) Mildenhall in July 1987 is Lynx AH.1 XZ680 of the LCF. Built in 1982, it is also known to have flown with 656 Sqn and, after conversion to AH.7 in 2005, 671 Sqn. By November 2011, it was being used for spares recovery at Air and Ground Aviation Ltd, Hixon, Staffordshire.

Lynx AH.1 ZD279 of 671 Sqn landing at Middle Wallop, July 1988. Built five years before, it was later converted to be an HAS.7 and is known to have also flown with 659 and 671 sqns. By 2013, its fuselage was at Air and Ground Aviation Ltd.

Seen at Hurn in June 1986 is a Lynx AH.1 XZ678, sporting the new camouflage scheme used by Germany-based AAC aircraft. This aircraft went to North Weald Heritage Aviation in Essex in 2021.

Photographed at RAF Mildenhall in May 1986 is Lynx AH.1 XZ182 of 3 Commando Brigade Air Sqn (CBAS), Royal Marines (RM). After the Falklands War, B Flt 3 CBAS received six Lynx AH.1s. In 1995, 3 CBAS became 847 Sqn. Built in 1978, XZ182 was converted to an AH.7 in 1989 but can now been seen in private hands outside a house at Bornem in Belgium.

Lynx HAS.2 XZ689 of 702 Sqn at Yeovilton in August 1978. Built the same year, it was later converted to be a HAS.3, then a HMA.8 and finally a HMA8.RSU. It is it currently being used for spares recovery at Hayward and Green Aviation Ltd.

Another view of Lynx HAS.2 XZ689 at Yeovilton in August 1980. Still with 702 Sqn, it was operating from HMS *Battleaxe*.

Also at Yeovilton in August 1980 was Lynx HAS.2 XZ255, also of 702 Sqn but operating from HMS *Cardiff*. It first flew in 1979, and it too became a HAS.3, HAS.3S, HMA.8 and HMA.8RSU. Its flying days over, it went into storage at RAF St Athan but is now believed to be at the Defence School of Electro-Mechanical Engineering (DSEME) in Lyneham, Wiltshire.

Lynx HAS.2 XZ252 first flew in November 1978 and was delivered the following year. Upgraded to HAS.3 standard, by 2007 it was being used for spares. This photograph was taken in August 1981 at Yeovilton, where the aircraft acquired the name 'Purdy'.

Lynx HAS.3 ZD250 of 702 Sqn is seen in August 1984. Built two years before, it also served with 829 and 815 sqns and was upgraded to a HAS.3S. By 2012, it was being used for ground instruction at the Centre of Aviation Medicine at RAF Henlow in Bedfordshire.

Lynx AH.1 XZ605 of 3 SBAS, Yeovilton, in September 1984. It first flew in November 1979, after which it flew with 651 Sqn, LCF and 3 CBAS. On 20 February 1989, it suffered a landing accident at Fagernes in Norway, but it was repaired at Fleetlands and upgraded to an AH.7. It then returned to 3 CBAS (847 Sqn) and spent the remainder of its service life with 671 Sqn. Its flying days ended in August 2013, and the following year it went to the Wattisham Station Heritage Museum in Suffolk.

Lynx AH.1 XZ203 of the LCF at Middle Wallop in July 1986. This aircraft also flew with 653, 671, 669 and 653 sqns, and during this time it was upgraded to AH.7 standard. By 2014, the cabin was being used for ground instruction at the Copehill Down Training Area, Salisbury Plain, Wiltshire.

Lynx AH.1 XZ205 of the Air Engineering Training Wg at Middle Wallop in July 1986. Converted to an AH.7, it now resides in the Old Hollow Paintball Park near Crawley in Sussex.

Lynx AH.1 ZD279 of 671 Sqn over Middle Wallop, July 1988. Built in 1983, it was upgraded to be an AH.7 but by 2013 was being used for spares recovery by Air and Ground Aviation Ltd.

Lynx AH.7 ZD560 of the Empire Test Pilot's School (ETPS) seen at Fairford in July 1989. It is now the gate guard at Yeovil Aerodrome in Somerset.

Chapter 2
Sea King

Above: Sea King HAS.1 XV642 first flew in 1969 and was upgraded to a HAS.2A. In 1997, it was being used for ground instruction at Aircraft Engineering and Survival School (AESS) Gosport. It was with HMS *Sultan* in 2023.

Right: Sea King HAS.2 XZ582 was delivered in 1977 and was later upgraded to HAS.5 standard. Seen here armed with Mk 47 torpedoes, it is coded 274 of 814 Sqn and operating from HMS *Hermes*. On 27 November 1989, it suffered a main gearbox failure en route to HMS *Invincible* and was ditched off Bermuda. At the time of the accident, it was coded 264 and flying with 814 Sqn.

An RNAS Culdrose-based Sea King HAS.1 of 706 Sqn. Apart from being an XV serial, the serial number cannot be made out for certain.

The folded rotor blades are obscuring the serial number of this Sea King HAS.1 of 706 Sqn, which was based at RNAS Culdrose. However, someone had painted the identifying flying fish badge of 820 Sqn on the door.

Above and below: Sea King HAS.1 XV644 was delivered to 700S Sqn in August 1969. Initially coded 582/CU, it became 528/PO. As 528/PO, it worked with 737 Sqn and, on 19 November 1974, suffered a tail rotor drive failure and ditched 21 miles south of Portland Bill. The remains were recovered to Lee-on-Solent, and by May 1994 it had gone to Predannack.

Seen at Lee-on-Solent in July 1972 is Sea King HAS.1 XV677. Delivered in January 1971, it flew with 820, 814, 819 and 825 sqns being converted over its career to HAS.2A, HAS.5 and HAS.6 standard. By 1999, it was a ground instructional airframe at Gosport but in 2006 moved to AeroVenture at the South Yorkshire Air Museum in Doncaster.

Photographed at RAF Chivenor in August 1971 is Sea King HAS.1 XV659 of 824 Sqn. Delivered the previous year, by 1978 it had been converted to a HAS.2A. It was with 824 Sqn for Operation *Corporate* and was later converted to HAS.5 standard by 1986 and HAS.6 by 1999. By 2001, it was a ground instruction airframe at HMS Sultan and then went to No.1 School of Technical Training at RAF Cosford. By 2022, it was reported as being back at HMS Sultan.

Sea King HAS.1 XV673 of 826 Sqn at RAF St Mawgan in August 1975. This aircraft first flew in September 1970 and was delivered the following month. It would be upgraded to HAS.2, HAS.5 and HU.5 standard. It is now on display at RNAS Culdrose with 771 Sqn markings.

The 826 Sqn badge is seen on the nose of Sea King HAS.1 XV673 at RAF St Mawgan in August 1975.

Sea King HAS.1 XV653 of 706 Sqn displaying at RNAS Yeovilton, September 1975. This aircraft first flew in November 1969 and was upgraded to HAS.2 standard in 1978, HAS.5 in 1983 and finally HAS.6. It is now a ground instruction aircraft at HMS Sultan.

Sea King HAS.1 XV653 of 706 Sqn, seen at RNAS Yeovilton exactly two years later.

Also seen at Yeovilton in September 1977 is Sea King HAS.1 XV657 of 706 Sqn. Delivered in February 1970, it was upgraded to HAS.2A, HAS.5 and HAS.6 standard. It was airlifted to Predannack in Cornwall by Chinook in June 2017 to be used for rescue training, and it was still there in 2022.

Sea King HAS.1 XV655 of 819 Sqn at RNAS Yeovilton in June 1977. It was also upgraded to HAS.2 and ultimately a HAS.6. It became a ground instruction airframe in 2001 and has resided at HMS Sultan since then.

Sea King HC.4 ZA290 of 846 Sqn, RNAS Yeovilton, August 1980. Delivered in 1979, on 18 May 1982, it was destroyed by its crew on the beach at Aguas Fresca, Chile, while landing Special Forces personnel.

Sea King HC.4s of 846 Sqn at RNAS Yeovilton in July 1982.

Sea King HC.4 ZA310 of 846 Sqn in July 1982. Delivered in September the previous year, it is believed to have been scrapped at Hitchin in 2014.

Seen at RAF Upper Heyford in August 1983 is Sea King HAS.5 of 814 Sqn. Delivered to the RN in May 1982, in 1993 it became a HU.5 and served with 771 Sqn, but in 2014 it became a spares 'Christmas Tree' for 771 Sqn's Sea Kings.

Above and below: Sea King AEW.2 XV656 of 849 Sqn is seen at RAF Mildenhall in May 1988. Built as a HAS.1 and delivered in February 1970, it was converted to a HAS.2 and in 1985 joined 849 Sqn as an AEW.2. By 2003, it had been converted to be a ASaC.7, after which it flew with 849, 854 and 857 sqns. It is currently believed to be in storage with HeliOps at Somerton in Somerset.

Getting airborne at RAF Mildenhall in May 1987 is Sea King AEW.2 XV650 of 849 Sqn. Built as a HAS.1 and delivered in November 1969, it was the second Sea King to be converted to AEW.2 standard. It would later become an ASaC.7, but on 22 March 2003, while operating from HMS *Ark Royal* in the Arabian Gulf, off Kuwait, it collided with Sea King XV704 and crashed into the sea. XV704 also crashed and sadly seven aircrew lost their lives.

An unusual sight at RAF Mildenhall in May 1988 – a Sea Harrier FRS.1 of 899 Sqn and a Sea King AEW.2 of 849 Sqn.

Sea King AEW.2 XV714 of 849 Sqn, RAF Mildenhall, May 1989. Originally a HAS.1, it became an AEW.2 and then ASaC.7. It ended its service with 849 Sqn and retired in September 2018, going into storage with HeliOps in Somerset.

Formation photograph of Sea King HAS.1s of 814 Sqn in July 1977. The nearest is HAS.1 XV675, which was built in 1970 and delivered to the RN in January 1971. It was later converted to HAS.2 and then HAS.6 standard, but by 2001 was being used for ground instruction at AESS, where it was still in 2022. The 'H' on the tails would indicate they were operating from HMS *Hermes*; 814 Sqn being on this carrier from 1973 to 1979 and XV675 being with 814 Sqn from September 1976 until transferring 819 Sqn in November 1977. It is believed to have force-landed off Portland on 2 August 1973 while with 737 Sqn.

Sea King HU.5 XV647 of 771 Sqn at RAF Upper Heyford in May 1990. Delivered as a HAS.1 in September 1969, it became a HAS.2, HAS.5 and then a HU.5 in 1987. It can be seen today at the Higher Purtington Showfield at Chard in Somerset.

Sea King HAS.5 XV700 of 705 Sqn, RAF Mildenhall, June 1984. Delivered as a HAS.1 in 1971, it became HAS.2, HAS.5, HAS.6, and finally HC.6(CR). On retirement, it was used for ground instruction at HMS Sultan and then at RAF St Mawgan in 2015.

Sea King HAS.2 XZ571 of 826 Sqn seen at RNAS Yeovilton in August 1978. At this time, it was serving on HMS *Tiger*, having been delivered two years before. Converted to HAS.2, HAS.5 and finally HAS.6 standard, it was sold to the Royal Australian Navy in November 2005. In 2011, it was reported as being in storage at Nowra, New South Wales.

Sea King HAS.5 ZA135 of 810 Sqn at RAF St Mawgan in August 1986. Delivered in July 1981, it was converted to HAS.6 standard and served with 849 Sqn. In 2012, it went to Air and Ground Aviation Ltd at Hixon, Stafford, for spares recovery, but the following year it was reported as being at a scrap yard in Oxfordshire. In 2022, its nose was reported as being seen at Beckley in Oxfordshire.

Chapter 3
Wasp/Scout

Westland Scout AH.1 XP849 was built in 1962 and flew with both the AAC and the Rotary Wing Sqn of the ETPS. On 3 April 1997, it suffered a heavy landing during auto-rotation training at Netheravon, near Amesbury, Wiltshire. It was sold in 2002, becoming G-CBUH, and then again to a private buyer in New Zealand in 2010 to become ZK-HQU. It is believed to currently be at Motueka, New Zealand, still in the ETPS white/red/blue colour scheme.

Above left, above right and below: Wasp HAS.1 XT442 displaying at an unknown airfield. Delivered in 1966, it first flew with 829 Sqn on HMS *Nubian*, then HMS *Zulu* and HMS *Yarmouth,* the latter between 1970–71. On 12 December 1973, it crashed on take-off from Norton Heliport, Dartmouth, while on loan from 829 Sqn (which was stationed on board HMS *Eskimo*) to the Britannia Royal Naval College. The main rotors hit the tail boom, after which the tail rotor shaft failed and the aircraft crashed and burst into flames. Sadly, all four on board, the pilot and three trainee officers, were killed.

Above and right: Wasp HAS.1 XT781 photographed at Biggin Hill in June 1971, being fitted with AS 12 missiles. Built in 1966, it first served with 845 Sqn, but by the time this photo was taken it was with 829 Sqn on HMS *Andromeda*. On 28 March 1974, it ditched off Portland but was recovered. It went into storage in 1980 and was sold to the Royal New Zealand Navy (RNZN) in 1983 as NZ3908. It was retired from naval service in 1998, when it was sold to Westland to become G-KAWW. In 2005, it was sold to South Africa to become ZU-HAS and was later known to be in private hands in Durban.

Wasp HAS.1 XT418 has quite a history. Delivered in 1964, it first flew with 829 Sqn then 703 Sqn. In January 1972, it ditched off Portland but was recovered, re-joining 829 Sqn on HMS *Endurance*. On 13 December 1981, it was about to land on the beach at St Andrews Bay, South Georgia, carrying the Governor of the Falkland Islands, Rex Hunt, and his wife. It landed heavily, bounced and went onto its side; luckily there were no crew or passenger injuries. It was then stripped for spares and abandoned.

Wasp HAS.1 XT426 of 829 Sqn photographed in March 1979. Delivered in May 1965, it flew with 706 Sqn and, following an accident on 28 September 1972 and subsequent repairs and time in storage, later flew with 829 Sqn. In 1992, it was delivered to the Royal Malaysian Navy, after which it went on display in the Maritime Museum in Malacca, Malaysia.

Seen at RNAS Yeovilton in August 1980 is Wasp HAS.1 XT421 of 829 Sqn, which at that time was serving on board HMS *Euryalus*. Built in 1964, it was sold to the Royal Malaysian Navy and, following its retirement, was used as an instructional airframe at the Malaysian Aviation Training Academy, Kuantan.

Wasp HAS.1 XT418 of 829 Sqn at Yeovilton in August 1981, seen here painted in accordance with operating from HMS *Endurance*. Four months later, it was written off in an accident on South Georgia.

Wasp HAS.1 XS543 of 703 Sqn at Middle Wallop in July 1982, armed with dummy AS.12 missiles. Delivered to the RN in 1964, shortly after this photo was taken, it was sold to the RNZN, and retired in 1995. It was then presented to the RNZAF Museum at Ohakea, and when that closed, it was moved to the new museum at Wigram.

An unidentified Wasp HAS.1, seen in the cold and wet Falkland Islands in 1983.

Wasp HAS.1 XT784 of 829 Sqn at RAF Greenham Common, possibly in 1988. It was sold to Malaysia the same year and after retirement in 2002 went on display at the Malaysia Air Force Museum in Kuala Lumpur.

Above and below: Seen at Yeovilton in September 1977 is Wasp HAS.1 XS567 of 829 Sqn, which was serving on HMS *Apollo* at that time. Built in 1964, it was apparently stored at Wroughton until joining 829 Sqn in 1970. Retired in 1992, it can now be seen on display at the Imperial War Museum, Duxford.

Wasp HAS.1 XT439 of 829 Sqn at Biggin Hill in June 1974. Built in 1965, this aircraft had a chequered history. On 14 July 1967, while flying with 829 Sqn on HMS *Ajax*, it ditched off Sembawang, Singapore, but was recovered, returned to the UK and repaired. Then, on 25 March 1986, still with 829 Sqn but operating from HMS *Rhyl*, it suffered a loss of control that saw the main rotor blades sever the tail rotor drive shaft. It crashed at Merryfield in Somerset and was written off. It is now in private hands in Hertfordshire.

Seen at RAF Lyneham in 1977 prior to air transportation is Wasp HAS.1 XT434 of 829 Sqn. Delivered in July 1965, it had a long career with 829 Sqn before being retired in 1988. By 1997, it was being used for ground instruction at Fleetlands in Gosport before being sold to Everett Aero in 2008. In 2009, it became G-CGGK and was owned by the Real Aeroplane Company at Breighton in Yorkshire. In 2022, it was sold privately.

Wasp HAS.1 XS542 of 829 Sqn seen at RAF Valley in 1973. Delivered in 1964, by 1978 it was in storage at Wroughton before being sold to Força Aeronaval de Marinha do Brasil (Brazilian Naval Aviation) and given the registration N-7042. On 6 December 1982, it crashed into the sea 80 miles off Natal on Brazil's northeastern tip.

Scout AH.1 XW613 is seen at Middle Wallop in 1980. Having been built in 1970, in January 1998, it joined the civil register as G-BXRS. It is now based at North Weald and is painted in the markings of 665 Sqn.

Above and below: Scout AH.1 XW280 displaying at Biggin Hill in May 1980. Built in 1969, at the time this photo was taken, it is believed it was being flown by the Development and Test Sqn at Middle Wallop, and it is armed with dummy SS.11 anti-tank guided missiles. This aircraft was scrapped at RAF Sek Kong in Hong Kong 1993.

Scout AH.1 XT624 of the Army Conversion Training Flt (ACTF) at RAF Greenham Common in May 1980. Built in 1966, it was noted as being with 660 Sqn at RAF Sek Kong in 1990. By 1997, it was on the civil register as G-NOTY.

Scout AH.1 XT642 of 664 Sqn at RAF Odiham in May 1977. This aircraft is known to have also flown with 656 Sqn in 1982, and was scrapped in 1993.

Above and below: Seen at Middle Wallop in August 1977 is Scout AH.1 XT636, which was flying with the Advanced Rotary Wing Flt (ARWF), which became the Rotary Continuation Squadron (RCS). Delivered in August 1966, it flew with 668 Sqn before becoming a ground instruction airframe in 1980. It was last seen preserved in a scrap yard at Kuala Belait, Brunei.

Seen overhead Middle Wallop is Scout AH.1 XR600 of the RCS. A very old aircraft, it was delivered in February 1964 and was reported as scrapped at Otterburn, Northumberland, in 1999.

Scout AH.1 XT632 is seen at Bassingbourn in May 1978. Built in 1966, it is known to have served with 658 Sqn but was auctioned by the Ministry of Defence (MoD) in 1995 and became G-BZBD. It crashed at Streatley in Berkshire on 23 August 2000, and its remains had been scrapped by 2007.

A pair of unidentified Scout AH.1s on the Falkland Islands in 1982.

Photographed in July 1985, Scout AH.1 XT831 was built in 1967. It was sold to GSM Helicopters in Ripon, North Yorkshire, in April 1977 and became G-BENG. On 9 September 1980, it crashed due to engine failure at Barmby Moor, East Yorkshire, while spraying pesticides on crops. It was written off in June 1983.

Chapter 4
Wessex

Wessex WL727 was the first Westland-built Wessex, designated as a WA.1 and then a HAS.1. It first flew in June 1958, after which it remained at Westland as a trials aircraft, especially for the HAS.3. It was also fitted with the engine and transmission/rotor assembly for the Napier Gazelle power plant and was at Rolls-Royce Hucknall from December 1965 to November 1967. It was last seen on the RAF Tern Hill dump in 1970 and had been burnt by 1975.

Wessex HU.5 XS479 first flew in November 1963, after which it flew with 700, 848, 772, 707 and 847 sqns. It went into storage in 1982, but two years later it was being used for ground instruction at the Joint Air Transport Establishment (JATE), RAF Brize Norton. It remained at JATE until 1996 and had been scrapped by 2000.

An unusual fate awaited Wessex HU.5 XT469. Delivered in 1965, it served with 848, 771 and 772 sqns as well as 847 Sqn (seen here) from 1969–1971. It went into storage in 1982, but four years later was being used for ground instruction at RAF Stafford. In 2009, it was sold to a scrap yard in Lancashire and in 2013 was on sale for £12,500 on eBay. In 2014, it was on sale by Ream Salvage for £11,000. The following year, it was reported as being at the Windmill Campersite on the Isle of Wight, where it was converted to be a unique camping experience; it is still there in 2023.

The nearest Wessex in this image is HU.5 XT454. It was delivered in June 1965, but on 31 October 1971, it was damaged by strong winds at RAF Luqa while with 845 Sqn. On 28 August 1972, it hit power lines and crashed at Bicton Wood, Cornwall, after which it was written off.

Above, right and below: Wessex HU.5 XS520 of 846 Sqn seen at Biggin Hill in June 1971. Built in 1964, it is fitted with two general-purpose machine guns, a drill SS 11 missile, 68mm rockets and flares for display only. It went to the School of Flight Deck Operations Fire School at Predannack in 2006 and was last seen at Predannack in 2017.

Wessex HU.5 XS479 was delivered to 700 Sqn in December 1963, after which it flew with 848 Sqn from 1964 to 1974. It is seen here operating with 848 Sqn off HMS *Albion*. It ended its days at JATE and was scrapped by 2000.

Wessex HU.5 XT461 of 846 Sqn at RNAS Yeovilton in July 1972. Delivered in 1965, it suffered engine failure with 771 Sqn and ditched into Mount's Bay, Cornwall on 16 October 1987.

Wessex HU.5 XT464, seen here with 846 Sqn, had quite a busy career. Delivered in 1965, it flew with 845 and 846 sqns. On 22 April 1982, it was en route to HMS *Antrim* from the Fortuna Glacier, South Georgia, carrying troopers from D Sqn Special Air Service when it crashed alongside XT473, also from 845 Sqn, in blizzard conditions.

The reasons why this Wessex HU.5 of 846 Sqn have been totally painted out is unknown. Photographed in June 1977, the code 'VL' could mean this was XT465 or XT451.

Above and below: Wessex HAS.1 XS868 of 737 Sqn is seen at Biggin Hill in September 1971 in its distinctive red/blue/red paint scheme. Delivered in 1965, it became the Fleetlands' gate guardian in 1981 but perished at the School of Flight Deck Operations Fire School by 2003.

Wessex HU.5 XT460 of 846 Sqn was delivered in July 1965, after which it flew with 856 and 846 sqns. It went to the Air Engineering School (AES) in 1986, then storage in 1987, and it was being used for battle damage repair training in 1991. It was scrapped in 1995.

Wessex XM872 was built as a HAS.1 and delivered in July 1961. In 1966, it was converted to HAS.3 standard and served with 819 Sqn (seen here at RAF Abingdon in 1969, coloured blue and yellow), 707 Sqn and finally, in 1976, 737 Sqn. On 11 March 1981, it suffered an engine failure off Portland and ditched off Durdle Door in Dorset.

Wessex HU.5s of 845 Sqn seen on Ascension Island, April 1983. In the background, a Heavy Lift Short Belfast C.1 G-BFTU taxies out.

Wessex HU.5 XT764 of 845 Sqn, again on Ascension Island in April 1983. This aircraft was delivered in December 1967, and in 1991 became a ground instruction airframe at Detmold in Germany. In 1995, it moved to Gütersloh and to HMS Sultan in 1999; it was scrapped at the latter in 2002.

Wessex HU.5 XS521 of 845 Sqn, Ascension Island, April 1983. This aircraft was delivered in January 1965 and is known to have flown with 707 and 772 sqns before coming a ground instruction airframe at Saighton Camp, near Chester, in 1987. Its remains had gone by 1998.

Wessex HU.5 XT771 of 772 Sqn is seen at RNAS Yeovilton in September 1984. Delivered in December 1967, it was destined to sail on the MV *Atlantic Conveyor* with 848 Sqn. However, needing repairs, it was replaced by XS512, which was lost when the ship was hit by an Exocet missile on 25 May 1982. XT771 went to the AES at Lee-on-Solent in 1988, then the RN Historic Flt in 2006 and the School of Flight Deck Operations, RNAS Culdrose, in 2014. It is currently at the Higher Purtington Showfield in Somerset.

Wessex HU.5 XS484 of 771 Sqn at RNAS Yeovilton in July 1982. Built in 1964, it spent much of its service career on trials, culminating with the Naval Aircraft Trials Installation Unit. In 1993, it was in storage at RAF Finningley and was then scrapped by Hanningford Metals in 1996.

Wessex HAS.3 XM870 of 737 Sqn, Yeovilton, July 1982. Built as a HAS.1 in 1961, it was converted to a HAS.3 in 1965. Shortly after this photograph was taken, it went into storage and was used for ground instruction at AES. In 1988, it moved to RNAS Culdrose, moving to Lee-on-Solent in 1993 and then HMS Sultan in 1998. It subsequently went to Predannack and was sold for scrap in 2007.

Above and below: Wessex HU.5 XT770 of 781 Sqn, Yeovilton, July 1982. Delivered in May 1967, for obvious reasons it was known as the 'Green Parrot'. It then went to 707 and 845 sqns before going into storage in 1985 and then to No. 1 School of Technical Training at RAF Halton in 1991. In 1994, it was at Strettons Outdoor Activity Centre, Shawell, being used on a paintball area, but in 2008 it became a submerged diving aid at the Stoney Cove Diving Centre, Leicester.

Seen at Farnborough in September 1982 is Wessex HU.5, being operated by the RAE. It joined 700 Sqn in 1964, after which it flew with 848 and 707 sqns. In 1968, it was being used by the A&AEE and as a de-icing test airframe for the RN. Later, it was with Naval Aircraft Trials Installation Unit. In 1986, it was being used as a ground instruction airframe by the RAE at Farnborough and in 1998 went to RAF Manston in Kent. It can now be seen at the Manston Museum.

Very little is known about Wessex HU.5 XS513, seen here with 847 Sqn at RNAS Yeovilton in July 1982. It was built in 1964 and also flew with 846, 781 and 845 sqns. In 2022, It was reported as dumped at Yeovilton.

Displaying at Yeovilton in July 1982 is Wessex HU.5 XT450 of 845 Sqn. Delivered in 1965, on 8 May 1982 it was flown to Ascension Island inside a Belfast C.1, and on 14 May 1982, it embarked on RFA *Tidespring* as a replacement aircraft in the Falkland Islands. On 10 July 1983, it was damaged in a heavy landing in Turkey, still with 845 Sqn, and the following year became a ground instructional aircraft at Predannack. It had perished by 1994.

Wessex HAS.3 XP156 of 737 Sqn seen at Yeovilton in September 1977. Built as an HAS.1 in 1962, it was believed to have been converted to an HAS.3 by 1968. On 27 June 1980, still with 737 Sqn, it crashed in the English Channel 16 miles off Portland Bill; three crew sadly losing their lives.

Above: Another Wessex HAS.1 converted to HAS.3 was XP118, seen here with 737 Sqn at RNAS Yeovilton in September 1976. On 15 July 1981 and still with 737 Sqn, it suffered an engine failure and ditched 15 miles south of Portland Bill; its crew were thankfully rescued. Its remains were recovered, and in 1982 it was headed for Predannack.

Left: A dramatic photograph of a 707 Sqn Wessex HU.5 seen at Yeovilton, September 1975.

Above and below: Wessex HU.5 XT466 of 771 Sqn at Keevil, Wiltshire, in June 1981. Built in 1965, by 1982 it was in storage at Wroughton. From 1987 to 1991, it was at No. 2 School of Technical Training at RAF Cosford, and in 1992 it was at Weeton Barracks, Preston. In 1997, it was reported as being at AES Gosport, then Whittington Barracks from 2009 to 2016, after which it became an exhibit at Morayvia, Kinloss, from 2018 and is still there today.

Wessex HU.5 XT760 of 846 Sqn, RNAS Yeovilton, August 1980. Built in 1966, it also served with 845, 847, 848 and 772 sqns. In 1991, it became a ground instruction airframe at the RN Engineering College in Manadon before going to AES and finally being sunk in Horsea Lake, Port Solent, Hampshire, for use by the Defence Diving School.

Wessex HU.5 XT453 of 845 Sqn, Yeovilton, August 1980. Built in 1965, it later flew with 781 Sqn. Its flying days over, it became a winch demonstrator but is today back at HMS Sultan.

Above and below: Formation landing of Wessex HU.5s of 707 Sqn led by XS489, Yeovilton, August 1980. Delivered to the RN in April 1964, very little is known about it apart from it was believed scrapped at Bromley in Kent.

RN Wessexes lined up at Yeovilton, August 1980. The nearest is HU.5 XS488 of 707 Sqn. Built in 1964, it crashed during a blizzard near Ose, Gratangen, Norway, on 30 March 1977 while with 845 Sqn. After repairs, it continued flying until going into storage at Wroughton in 1982. It was then used for ground instruction at No. 1 School of Technical Training, RAF Halton, Wattisham Air Base and HMS Sultan. It was sold privately in 2013 and then went to Gunsmoke Paintball at Layer Wood, Tiptree. It is now submerged in Cromhall Quarry, Gloucestershire.

Seen onboard HMS *Ark Royal* in the Caribbean in 1978 is Wessex HAS.1 XP151 of 771 Sqn. After a career lasting from 1962 to 1979, it went into storage at Wroughton. It was later used for fire training at Predannack and had perished by 1997.

Chapter 5
Whirlwind

A very old photograph of a new Whirlwind HAR.1. Although the Royal Navy wording and the serial stating XA give some clues, it is hard to tell whether this aircraft is XA842 or one from the batch XA862 to XA856.

Whirlwind HAS.7 XN302 first flew in November 1959, after which it flew with 848 and 847 sqns. By 1975, it was a ground instructional aircraft as AES, having flown from RNAS Culdrose to RNAS Lee-on-Solent on 13 April 1975. By 1983, it was outside the fire station at Lee-on-Solent, but in 1987 it was being used for ground instruction at HMS Royal Arthur. Three years later, it was at RAF Finningley, but, three years after that, it was reported as being on the dump. It was believed to have been scrapped in 1996.

Whirlwind XN309 was built as a HAS.7 in 1960 but was later converted to a HAR.9. On 19 December 1962, with 846 Sqn, it damaged its tail rotor landing in a clearing near Limbang, Malaysia. By January 1968, it was with the Culdrose SAR Flt and was still with the Flt in January 1975 when this photograph was taken. In 1975, it was being used for ground instruction. It then went to the Second World War Aircraft Preservation Society at Lasham, Hampshire, which was dissolved in 2010. However, it is thought that, in 1992, XN309 went to Baldonnel near Dublin as a fire/rescue training airframe and was later scrapped.

Seen at RNAS Lee-on-Solent in July 1972 is Whirlwind HAS.7 XN259. Between 1959 and 1969, it flew with 42 Commando and 848 Sqn. It was with 771 Sqn on 26 July 1969 when it suffered engine problems when in the hover at RNAS Culdrose, and was damaged in the force-landing. It then became a ground instruction airframe first at Arbroath and later Lee-on-Solent. It went to the Proof and Experimental Establishment (P&EE) at Shoeburyness in 1979 and by 1990 it was on the fire dump at London City Airport. It was scrapped seven years later.

Another Whirlwind HAS.7 at Lee-on-Solent in July 1972 was XA870. First flown in May 1954, it then flew with 848 Sqn and the RAF's 155 Sqn in the Far East until 1955. From 1955 until 1966, it was part of 701 Sqn's HMS *Protector* Flt (and its penguin badge would have been on the nose of this blue and red aircraft). From 1966 until 1973, it was an instructional airframe at Lee-on-Solent. In 1976, it was at the Cornwall Aeronautical Park and in 2002 was donated to the Yorkshire Helicopter Preservation Group. It can now be seen at AeroVenture.

Close-up of the winch on an unidentified RN Whirlwind.

Out with the old, in with the new. Lynx HAS.2 XX510, the Lynx prototype that first flew in March 1973, leads two Whirlwind HAS.7s of the Joint Helicopter Tactical Development Unit, part of the Joint Warfare Establishment (JWE), over Old Sarum in Wiltshire. The nearest aircraft is XN299, which first flew in October 1959. It then flew with 848, 847 and 845 sqns before joining the JWE in 1967. In 1976, it went into storage, but three years later was at the Torbay Aircraft Museum (which closed in 1988). Since 1997, it has been at the Tangmere Military Aviation Museum in West Sussex.

Photographed in October 1975 is Whirlwind HAR.3 XG577. Delivered to the RN in March 1956, it is known to have flown with 701, 705, 815 and 837 sqns before becoming a ground instruction airframe at Arbroath in 1967. It then went to Lee-on-Solent and from 1974 was displayed at a number of museums before going into storage at the Imperial War Museum, Duxford, in 1989. The following year, it was seen on RAF Leconfield's fire dump before being finally scrapped in 2001.

Above and right: Seen at Royal Naval Aircraft Yard (RNAY) Wroughton in June 1976 is Whirlwind HAR.9 XL839. Built as a HAS.7 in 1957 and upgraded to a HAR.9 in 1964, its last unit was RNAS Culdrose's SAR Flt (hence 'Seahawk' on the nose). It went into storage at Wroughton in 1975 but was scrapped at Lee-on-Solent in 1983.

Seen at Biggin Hill in June 1971 is Whirlwind HAS.7 XM660. Built in 1958, by 1974 it was being used as a ground instruction airframe for RN recruiting purposes. It then became a gate guard at Fleetlands and RNAY Almondbank before going to the North East Aircraft Museum (now the North East Land, Sea and Air Museum) in 1984. It was then at the RAF Millom Aviation & Military Museum from 1996 to 2010, after which it was shipped to Europe. Until 2019, it was at Bruges in Belgium but since then is believed to be at Koropi Kalivia, Greece.

Photographed at RNAS Lee-on-Solent in July 1972, the serial number of this Whirlwind HAS.7 is not visible, but the 'PO' for Portland at 518 would indicate it is possibly from 829 Sqn; the HAS.7 serving with this squadron from 1965 to 1967.

Above, right and below: Whirlwind HAR.9 XL898 is seen at RAF Abingdon in September 1980. Built as a HAS.7 in 1958, it flew with 824, 825, 848, 847 sqns before being converted to a HAR.9 in 1966. It was then modified for Antarctic operations and flew with 829 Sqn from HMS *Protector* and HMS *Endurance,* hence the penguin badge favoured by the squadron. By 1978, it was in storage at Wroughton. It was last seen on the dump at Boscombe Down in the early 1990s.

Above and below: Photographed at RNAS Lee-on-Solent in July 1971 is Whirlwind HAR.3 XG574. It first flew in August 1955 but by 1967 was a ground instruction airframe at RNAS Lee-on-Solent, after which it went into storage at RNAY Wroughton in 1975. It is currently on display in Hall 1 of the FAA Museum, Yeovil.

Seen in a very poor state at the South Wales Air Museum at Rhoose, Glamorgan, in September 1978 is Whirlwind HAS.7 XG592. Delivered to the RN in 1957, it flew with 700, 705 and 846 sqns. Its final squadron was 705 Sqn, and it left RNAS Culdrose for storage at Wroughton in 1974. It then went to South Wales Aircraft Museum, but by 1995 it was at Taskforce Paintball at Cowbridge in the Vale of Glamorgan. It is still there in 2023, albeit in a very poor state.

In a poor state at Boscombe Down is Whirlwind HAR.9 XL898. Still visible on the nose is the 829 Sqn HMS *Endurance* penguin badge. Shortly after this photograph was taken, the remains were disposed of.

Seen at the Newark Air Museum in July 1991 is Whirlwind HAS.7 XM685. It first flew in February 1959 but was sold to Autair at Luton in 1971 with the registration G-AYZJ. After changing ownership, it appears to have gone into storage in 1976, only to be obtained by the Museum in 1980 where it is still in 2023 in the markings of 771 Sqn.

Chapter 6
Minor Types

Above: Seen here is Aerospatiale Alouette AH.2 XR 386. Two were purchased in 1958 for trials with the Army to work as liaison helicopters, carrying passengers and light tactical stores around the battlefield. Fifteen more were ordered in 1961 when it became clear that the Westland Scout would take longer than planned to come into service. It had reasonable hot and high performance and was very reliable technically, proving successful with 16 Flt and United Nations Peacekeeping Force in Cyprus (UNFICYP). This aircraft was built in 1961 and is known to have served with 10 Flt at Dhekelia in Cyprus, 24 Flt at Detmold in Germany and UNFICYP. The Alouette was replaced by the Gazelle AH1 in 1988, and on 9 March 1990, the remaining fleet (bar two) was offered for sale.

Opposite: Aerospatiale/Westland Gazelle AH.1 XW843 first flew in January 1972 and was delivered in October the same year. On 25 February 1990, this aircraft, along with four others (including XZ348) operating with 2 Flt, Allied Command Europe Mobile Force, was hover-taxiing in for a refuel when the two front Gazelles clipped each other, causing bits of rotor blades to fly off. The debris hit the third Gazelle and the fourth over torqued trying to avoid the debris. The fifth aircraft was far enough back not to be involved. XW843 and XZ348 received Cat 5 damage, while the other two were Cat 4 and Cat 3. After being written off, XW843 was scrapped at Arborfield, Berkshire, in 1996.

Above and below: Seen here are the red and white-coloured Gazelle HT.2s of the RN's Sharks helicopter display team. Formed in 1975 and flown by instructors from 705 Sqn at RNAS Culdrose, the team initially had six aircraft but following an accident in 1977, the team was reduced to four aircraft. The Sharks was disbanded in 1992. The first photograph shows XW858/50 and XW895/51. Both joined 705 Sqn in 1979, the former going into storage in 1997 to be sold in 1999 to become G-CBKD, while the latter was also sold to become G-BXZD after which it went to Ukraine. The second photo shows XW884/41 (right) and XW894/52. The former was sold in 2001, went to Swaziland that year as 3D-HXL, and later returned to the UK to become G-BZDV, after which it was owned by a number of companies and individuals until it landed in private hands in the Ashurst area. XW894 was delivered to 705 Sqn in 1974, then withdrawn from service in 1997 and sold off in 2000 to become G-BZOS, but on 14 July 2002, it crashed at Gaydon in Warwickshire, after which it was scrapped.

Included even though they are not Army or RN are these RAF Bell Sioux HT.2s; the only serial visible is XV323 on the top aircraft. It was delivered in March 1968 and appears to have spent much of its time with the Central Flying School at RAF Ternhill. It was sold to British Executive Air Service with registration G-BCZL in 1975. It was written off on 25 July 1977, following a crash at Bishops Norton, Lincolnshire.

Although retired from the AAC in 1967, this Westland Skeeter AOP.12 XM553 was photographed at Middle Wallop in August 1977. This type was used in reconnaissance, air observation, liaison and training roles. XM553 was built in 1959 and on retirement went on the civil register as G-AWSV. It can now be seen on display at the Yorkshire Air Museum at Elvington, near York.

Seen at RAF St Athan in September 1977 is Skeeter AOP.12 XN341. Delivered to 651 Sqn in February 1960, it became a ground instruction airframe at RAF St Athan in 1968 and was sold in 1989. In 2000, it was displayed at the Stondon Transport Museum in Bedfordshire but was sold on eBay in 2015, after which it appeared on stands on a roof in Maasbracht, Holland.

Skeeter AOP.12 XM553 getting airborne at Middle Wallop in August 1977.

Skeeter AOP.12 XL814 of the AAC Historic Flt at RAF Mildenhall in May 1989. Delivered in March 1959, it is now preserved at the Army Flying Museum at Middle Wallop.

Skeeter AOP.12 XL814 seen at RAF Upper Heyford in May 1990.

Skeeter AOP.12 XL814 photographed at Middle Wallop, date unknown.

Above and below: The Army helicopter display team, the Blue Eagles, was formed in 1968 and initially flew the Sioux HT.2. It disbanded in 1974.

Sioux AH.1 XT202 of 653 Sqn is seen at Greenham Common in August 1976. Built in 1966, it was sold to Germany to become D-HEEE and then to Spain to become EC-DVM.

Photographed at RAF Lyneham in February 1977 is Sioux AH.1 XT841. Built in 1967, it was sold in 1978 to become G-BFJT and was later owned by GSM Helicopters at Ripon, North Yorkshire, until it was sold to a buyer in Cyprus in 1982 to become 5B-CEV. It has since been broken up.

Another Sioux AH.1 seen at RAF Lyneham in February 1977 was XT555. Also built in 1967, it was sold to Germany to become D-HAFS and was last heard of as being at Rue Neuve, Rhone-Alpes, France.

Photographed in May 1977 at Sioux are AH.1 XT511 and XT130. Both flew with the Blue Eagles in their careers; the former was then sold to a buyer in South Africa, while the latter was sold to a buyer in Canada in 1977.

Left and below: Sioux AH.1 XT153 is at Greenham Common in June 1977, seen here with 14 Flt. It too flew with the Blue Eagles. Built in 1965, after its time with the AAC, it was sold to a private buyer in Germany to become D-HOFD. It was then privately sold to South Africa, only to be written off on 7 December 1981.

Above and below: Sioux AH.1 XT108 of the ARWF at Middle Wallop in August 1977. Built in 1965, it became the Middle Wallop gate guard in 1979, but since 1998 it has been displayed at The Army Museum of Flying in Middle Wallop.

Above, left and below: Sioux AH.1 XT131 seen at RAF St Athan in August 1978. Built in 1964, it is seen here with the ARWF. It went into storage in 1978 but in 1979 was flying with the Army Historic Aircraft Flt. Funding was withdrawn from the Flt in 2013, and by 2015, it was with the Historic Army Aircraft Flt Trustee Ltd with the registration G-CICN. It is still flying with the trust today.

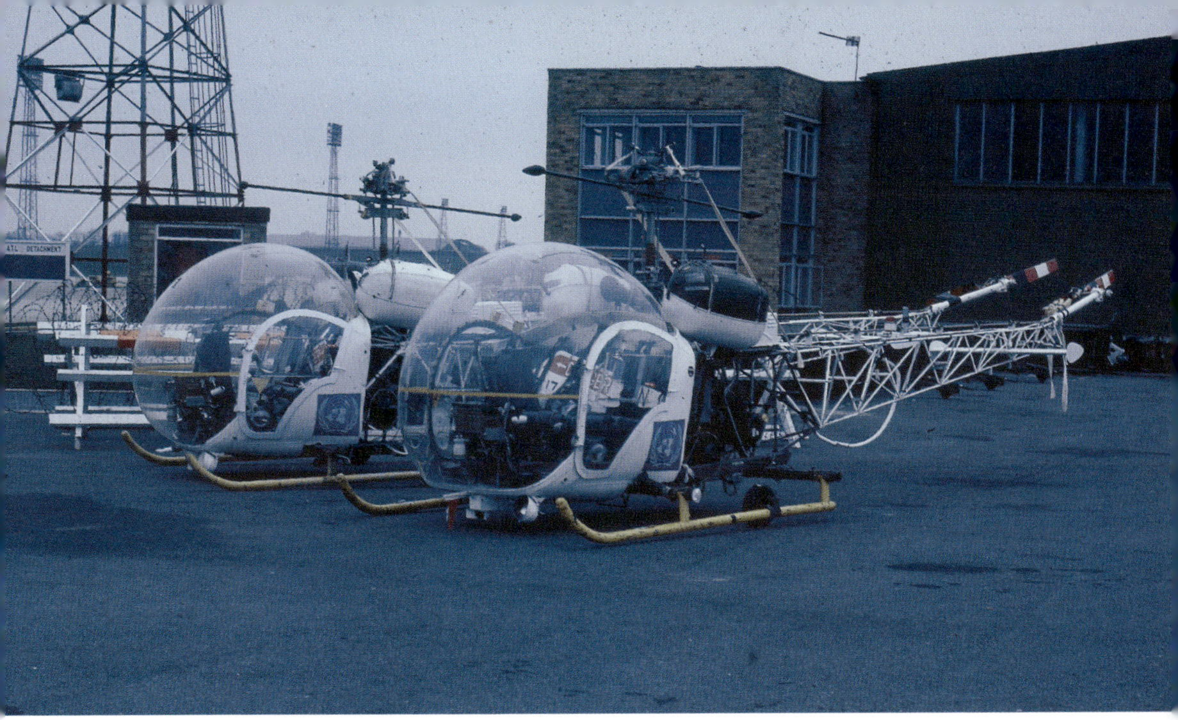

Photographed at RAF Lyneham in March 1978 are Sioux AH.1 XT190 and XT170, both in United Nations's markings. The former was built in 1965, and shortly after this photograph was taken, went into storage at RNAY Wroughton. In 1993, it became the gate guard at Soest army base in Germany but three years later moved to Wattisham in Suffolk. It is now on display at the Helicopter Museum in Weston-super-Mare. XT179 was sold privately later in 1978 and was initially located at Doncaster and then Heston in Middlesex with the registration G-BFYF. In 1985, it was sold to a buyer in Australia, becoming VH-HMK. It was then converted to be a Westland-Bell-Soloy 47G-3B1 and flown by McDermott Aviation/Heli-Lift Australia at Cooroy, Queensland, and, according to the company website, is still being used for the eradication of red fire ants, yellow crazy ants and the control of Mosquito larvae.

Gazelle AH.1 XW847 was built in 1973, after which it seems to have spent much of its time with 665 Sqn at RAF Aldergrove, Northern Ireland. It is still airworthy today, now flying with 7 (Training) Regt Army Air Corps from Middle Wallop.

Above and below: Seen at RNAS Yeovilton in September 1975 is Gazelle HT.2 XW856 of 705 Sqn. Built in 1973, it was sold and became G-CBBY and later G-GAZL. In 2006, it was later reported as being in private hands in the Ukraine.

Seen at Greenham Common in August 1976 is Gazelle AH.1 XX379, at that time with 2 Flt, AAC. Built in 1974, by 2013 it was being stored at RAF Shawbury. It was last reported as being used for instruction at Boscombe Down.

The 2 Flt badge on the nose of Gazelle AH.1 XX379. 2 Flt had been formed at Netheravon in Wiltshire in 1974.

2 Flt was part of Allied Mobile Force (AMF), and because of this XX379 carried the AMF shield on its nose.

Gazelle AH.1 XX411 of 3 CBAS at RAF Lyneham in February 1977. On 21 May 1982, Sgts Andy Evans RM and Eddy Candlish RM were providing an armed escort to a Sea King headed towards Port San Carlos, Falkland Islands, when they were hit by small arms fire and ditched into San Carlos Water. Sgt Candlish managed to tow the wounded Sgt Evans to shore, but the latter later died of his wounds. Sgt Candlish received a Mentioned in Despatches for his actions that day.

Gazelle AH.1 XX412 of 3 CBAS at Middle Wallop in July 1982. Built in 1975, in 1980 it was with the RM at Arbroath as M Flt, 3 CBAS. On 5 April 1982, it embarked on RFA *Sir Galahad*, headed for the Falkland Islands. 3 CBAS then became 847 Sqn, but by 2012 XX412 was at RAF Cosford, where it is still today.

Gazelle AH.1 XZ315 of the ARW Flt getting airborne from Middle Wallop. Built in 1977, it also flew with 665 Sqn at RAF Aldergrove. By 2010, it was reported as being partly scrapped.

Gazelle AH.1 XW888 of ARWF at Middle Wallop. Built in 1974, it is thought to have served with to ARW Flt. It was then used for ground instruction at the Defence College of Aeronautical Engineering at Arborfield in Berkshire, and then went to the Defence School of Electro-Mechanical Engineering at Lyneham in Wiltshire. In 2018, it was sold by auction by Agility DGS Ltd and is now in private hands at East Hanney, Oxfordshire.

Gazelle HT.2 XX391 of 705 Sqn at Greenham Common in May 1980. Built in 1975, in 2001 it had been sold to Christchurch in New Zealand with the registration ZK-HTB. This registration was cancelled in 2010 and it is thought to have been exported to Russia.

The 705 Sqn badge seen on the tail of Gazelle HT.2 XW907 at Greenham Common in May 1980. Built in 1974, XW907 had been sold by 2000 to South West Aviation to become G-BZOT. A series of private owners followed and in 2002 it was re-registered G-DFKI.

Being small and able to fly low when the weather permitted, the Gazelle AH.1, seen here in the Falkland Islands in 1982 and believed to be from 3 CBAS, was a valued asset to both the Army and Royal Marines.

Appendix
Variants

The following variants are mentioned in this book:

Lynx
WG.13	Prototype that first flew in March 1971; 13 built.
AH.1	Initial production version for British Army and used for tactical transport, armed escort, anti-tank, reconnaissance and casualty evacuation.
AH.7	Upgraded version with Gem 41-1 engines, upgraded gearbox, new composite material tail rotor; 12 new-builds and 107 AH.1s converted.
HAS.2	Initial production version for the RN.
HAS.3	Improved version of HAS.2 with Gem 42-1 engine and upgraded gearbox; 30 new-builds with 53 HAS.2 converted to HAS.3.
HAS.3S	Improved version of HAS.3 fitted with secure comms.
HMA.8	Upgraded maritime attack version based on Super Lynx 100.
HMA.8 DSP	Fitted with Digital Signal Processor.
HMA.8.SRU	Defensive Aid Subsystem aircraft modified with Saturn Radio Upgrade.

Sea King
AEW.2	Conversion of Sea King HAS.1 or HAS.2 for AEW to fill gap revealed during the Falklands Conflict.
ASaC.7	Upgraded AEW.2/5 with Searchwater 2000AEW radar.
HAS.1	First ASW version for RN.
HAS.2A	Upgraded HAS.1 with Gnome H.1400-1 engines.
HAS.5	Upgraded ASW version with longer range, Super Searcher radar and improved acoustic-processing systems.
HAS.6	Upgraded ASW version with improved avionics, new sonar processor, improved tactical displays and better comms.
HC.4	Commando assault and transportation.
HC.6 (CR)	Surplus HAS.6 ASW aircraft converted for commando assault.
HU.5	Surplus HAS.5 ASW aircraft converted for utility role.

Wasp/Scout
Scout AH.1	Light utility helicopter for the British Army.
Wasp HAS.1	Shipboard ASW helicopter for the RN.

Wessex
WA.1	Prototype.
HAS.1	RN utility, ASW and ASR.
HAS.3	Improved HAS.1 with new avionics. Three new-build and 43 converted from HAS.1.
HU.5	Troop transport.

Whirlwind

HAR.1	RN SAR; 10 built.
HAR.3	RN SAR; 25 built. Wright R-1300 Cyclone 7 engine.
HAR.9	12 converted HAS.7 or HAR.7 with Bristol Siddeley Gnome engine.
HAS.7	RN ASW and commando transport.

Minor Types

Alouette AH.2	Liaison, passenger and stores transport.
Gazelle AH.1	British Army and Royal Marines (RM) light transport, scouting and attack.
Gazelle HT.2	RN Training.
Sioux AH.1	General-purpose helicopter for British Army and RM.
Sioux HT.2	RAF Training.
Skeeter AOP.12	Reconnaissance, air observation, liaison and training for British Army.

Other books you might like:

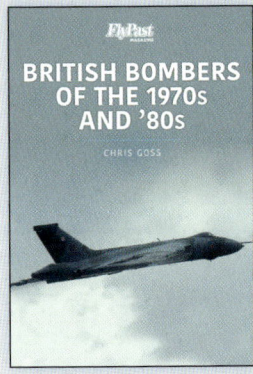

BRITISH BOMBERS OF THE 1970s AND '80s
CHRIS GOSS

Historic Military Aircraft Series, Vol. 4

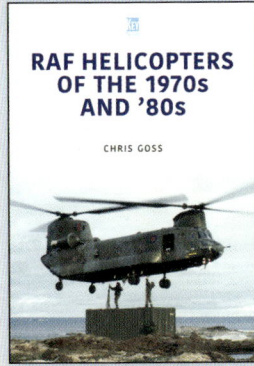

RAF HELICOPTERS OF THE 1970s AND '80s
CHRIS GOSS

Historic Military Aircraft Series, Vol. 18

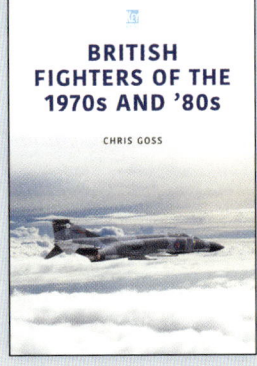

BRITISH FIGHTERS OF THE 1970s AND '80s
CHRIS GOSS

Historic Military Aircraft Series, Vol. 2

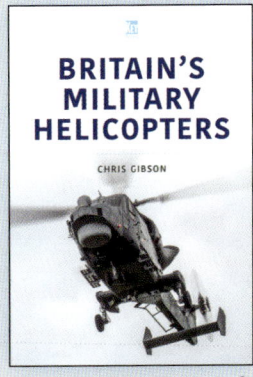

BRITAIN'S MILITARY HELICOPTERS
CHRIS GIBSON

Modern Military Aircraft Series, Vol. 4

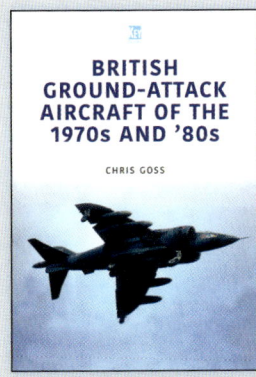

BRITISH GROUND-ATTACK AIRCRAFT OF THE 1970s AND '80s
CHRIS GOSS

Historic Military Aircraft Series, Vol. 8

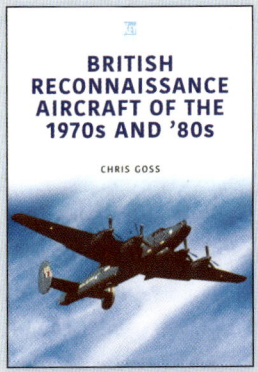

BRITISH RECONNAISSANCE AIRCRAFT OF THE 1970s AND '80s
CHRIS GOSS

Historic Military Aircraft Series, Vol. 10

For our full range of titles please visit:
shop.keypublishing.com/books

VIP Book Club

Sign up today and receive
TWO FREE E-BOOKS

Be the first to find out about our forthcoming book releases and receive exclusive offers.

Register now at **keypublishing.com/vip-book-club**

Our VIP Book Club is a 100% spam-free zone, and we will never share your email with anyone else. You can read our full privacy policy at: privacy.keypublishing.com